The Mosquito Book

by Scott Anderson & Tony Dierckins

Dennoch Press

Published by Dennoch Press • 808 Martha Street • Duluth, Minnesota 55805

Distributed by Adventure Publications • 800-678-7006

The Mosquito Book

Text: Tony Dierckins and Scott Anderson
Research assistance: Laura Witrak
Illustrations: Mike Pearce
Cover design: Jeff Brownell

Printed in the United States of America by LithoColor Press

10 9 8 7 6 5 4 3 2 1

ISBN: 0-9644521-1-1

Library of Congress Cataloging in Publication Data: 98-092515

No mosquitoes were harmed in the production of this book. Sorry.

Special Thanks to...

Dr. Richard Anderson, U.S. Environmental Protection Agency entomologist, for verifying the accuracy of our information;

Dr. Paul Anderson, biochemist, for the dinosaur DNA data;

and Dr. Mary Sue Lux, veterinarian, for her help on questions concerning nonhuman mosquito victims.

You'd better read this:

We're from Minnesota, a place so thick with mosquitoes that many people believe it is the state bird. We are not, however, experts on mosquitoes. We're just a couple of guys who've spent too much time waving our hands in the air trying to shoo away the annoying little bloodsuckers. We hate them.

Our hatred of mosquitoes prompted us to do some research. After years of blindly lashing out against the mini-blood banks that have made our summers unbearable, we decided to educate ourselves—to find out as much as we could about mosquitoes in order to deal with them as effectively as possible. What we found changed our lives, and we think it will do the same for you. That's why we wrote it all out in this book.

If, however, you use the information in this book incorrectly and somehow injure yourself or your loved ones, don't come to us with your fancy lawyers and their high-falutin' lawsuits. Even though we found a real card-carrying bug expert to verify our mosquito facts, this book is NOT a scientific textbook; it's just a fun little book filled with interesting stuff about mosquitoes. We hope you enjoy it.

— The Authors

Contents

*Additional "Mosquito Miscellany" has been disbursed randomly throughout this book. You'll know you stumbled across it when you see this happy little fellow at the left. Try not to let this confuse you.

mosquito life
Sex, blood, and standing water

WHAT EXACTLY IS A MOSQUITO?

According to *Merriam Webster's Collegiate Dictionary*, a mosquito is "any of a family (Culicidae) of dipterian flies with females that have a set of slender organs in the proboscis adapted to puncture the skin of animals and to suck their blood."

You know those bugs that look like great big huge mosquitoes and fly slow and lazy but never seem to try to take a bite out of you? They're not mosquitoes! They're flies of the family Tipulidae. Some people call them daddy longlegs.*

**Of course the moniker "daddy longlegs" is also used to describe some big-'ol spiders.*

MOSQUITOES DON'T NEED YOUR BLOOD FOR FOOD.

Contrary to popular belief, mosquitoes don't get their nutrition from your blood.

Their nutritional needs are primarily met by flower nectar.

SO WHY DO MOSQUITOES NEED MY BLOOD?

Blood provides the protein mosquito eggs need for development.

While fertilization occurs only after a blood meal,
the blood does not fertilize the eggs.*

***As with humans, the male is necessary for at least one thing.**

ONLY THE FEMALE MOSQUITO "BITES."

Yes, but you probably already knew that lady mosquitoes are the bloodsuckers of the family.

Due to years of sensitivity training, the authors refrain from further comment on this issue.

How the...?

MOSQUITOES DON'T "SUCK" YOUR BLOOD.

They pump it!

The female mosquito has a pump in her head which she uses to pull in the blood. It works a lot like a turkey baster.*

**Even though they don't technically suck blood, we're still going to refer to the little bloodsuckers as "bloodsuckers."*

HOW MUCH BLOOD IS LOST FROM A MOSQUITO "BITE"?

The average mosquito consumes one millionth of a gallon of blood per "bite."

At that rate, it would take about 1,120,000 bites to drain the blood from an average adult human.

HOW DO MOSQUITOES ACTUALLY "BITE"?

They don't bite: It's more like they go in and rip things up a bit.

The female mosquito's mouth is composed of six long piercing parts called stylets. Once inserted into the skin (1), four of the stylets—armed with serrated edges—saw back and forth, ripping apart tiny capillaries just beneath the skin's surface to create a pool of blood (2). The fifth stylet injects saliva, which acts as an anticoagulant to help the blood flow smoothly (3). The last stylet is shaped like a trough. When the others wrap around it, they form a tube through which the blood is pumped (4). The mosquito then enjoys a little drink of blood (5).*

***If the mosquito is lucky, it will hit a vein and pump from there.**

HOW MOSQUITOES BITE

(SEE TEXT ON FACING PAGE)

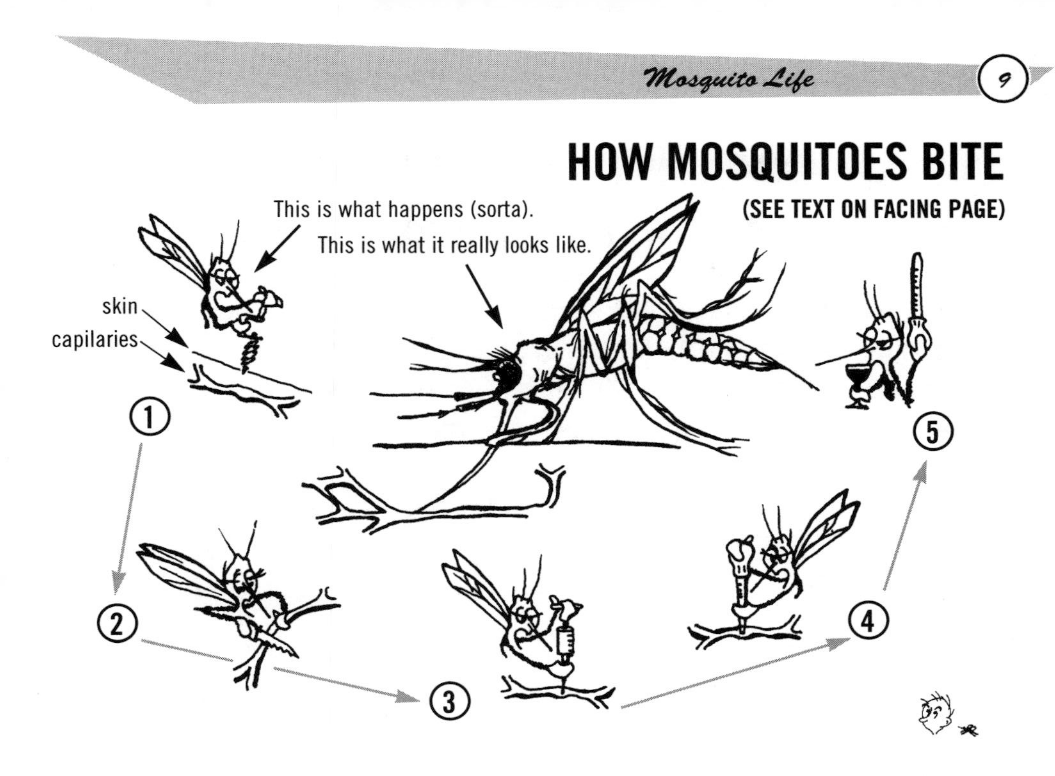

HOW MANY TIMES DOES A MOSQUITO BITE?

Some people think mosquitoes, like bees, sting once and die. But remember, mosquitoes are taking blood, not defending their queen. If they died after "stinging" once, it would defeat the purpose of taking the blood in the first place.

A female mosquito goes out for blood whenever she needs protein for her eggs. She can feed multiple times and usually makes between one and three batches of eggs during her lifetime.*

**Unless, of course, she gets squashed on her first attempt.*

IF BITTEN BY A RADIOACTIVE MOSQUITO, DO YOU ACQUIRE SUPERPOWERS THAT ALLOW YOU TO FIGHT CRIME AND EVIL?

No.

You don't get your own comic strip either.

Now stop asking us silly questions so we can get on with the book.

DO MOSQUITOES TAKE BLOOD ONLY FROM HUMANS?

No! They'll take blood from just about anything with blood: mammals, birds, reptiles, etc.

According to Will Barker, author of *Familiar Insects of North America*, the female mosquito can puncture many types of body covering—even the leathery skin of a frog or the overlapping scales on a snake.*

**And we're surprised when they get through a cotton T-shirt.*

Say it ain't so!

THEY EAT THEIR OWN!

While mosquitoes don't "bite" other mosquitoes, they sometimes go cannibalistic if crowded in the larval stage, where larger, older larvae eat the smaller, freshly hatched larvae.

Some mosquito species even prey on other mosquito species. In fact, scientists are working on methods to produce cannibalistic mosquito larvae as a population control method.

CAN MOSQUITOES "EAT" TOO MUCH?

Not if everything's working right. If a mosquito gets too bloated with blood to fly away from her victim, she releases a little ballast to help her become airborne.

How?

She empties the mosquito equivalent to a bladder.*

*In other words, she piddles on you. Icky.

HOW DO MOSQUITOES KNOW WHEN TO STOP FEEDING?

The female stops feeding once stretch receptors in her abdomen have been triggered.

Experiments have shown that if the nerves connecting the stretch receptors to the brain are cut, the female will take blood until she bursts!

Fun with mosquitoes!

TURNING THE TABLES

The next time a mosquito lands on your forearm, don't swat it. Instead, pinch your skin on either side so that the pressure traps her stinger in your arm—but not so tight that you cut off the bloodstream. Even when her sensors tell her it's time to stop, she'll keep taking in blood—until she explodes!*

***We've never actually tried this. It may be a myth.**

HOW DO MOSQUITOES ATTRACT A MATE?

Male mosquitoes—swarms of them—are attracted to a female by the whine given off by her beating wings. Females' wings beat slower, and have a lower pitch, than males'. (The average wing speed of mosquitoes varies between 250 and 500 beats per second.)

In fact, scientists are able to use tuning forks set at the pitch of a female to attract (and ensnare) males.

However, if you're not a mosquito, whining probably isn't the best way to attract a mate.

DO MOSQUITOES HAVE A MATING RITUAL?

No. Mosquitoes mate after the female flies into a swarm of males—a huge mosquito singles bar if you will.

This swarm may be as small as a softball or as large as a classroom, and mating takes place almost immediately—in midair! The happy couple eventually floats to the ground.*

*Mating takes anywhere from four to forty seconds, but some couples have been known to stay together for over an hour.

Don't try this at home!

BORN BACKWARDS

After emerging from the pupal skin, the male mosquito is not attracted to females.

A good thing, too: when a male mosquito emerges, his sex organs are backwards, making mating impossible. This state lasts for about a day (depending on the species and the temperature). The terminal segments of the abdomen that hold the sex organs then rotate 180 degrees, and he's good to go.

Why do they know this?

STRANGE BUT TRUE!

In scientific experiments, male mosquitoes continued to copulate even after being decapitated!*

**If you ask us, some scientists have way too much free time on their hands.*

HOW MANY EGGS DOES A MOSQUITO LAY PER "BATCH"?

Lots!

Depending on the species and how much blood she has consumed, a female mosquito can lay as many as several hundred eggs in one batch.

HOW MANY TIMES DOES A FEMALE NEED TO MATE FOR EACH BATCH OF EGGS SHE LAYS?

She needs to mate only once before laying many batches of eggs.*

**And after she has the sperm she needs to reproduce, a female will fight off further attempts at mating.*

WHERE DO MOSQUITOES LAY THEIR EGGS?

Many mosquitoes lay their eggs in standing water. Floodwater mosquitoes (*Aedes*) lay eggs at the edge of water, the water recedes, and the eggs lie dormant until the water rises again.

The *Culex* lay their eggs on the water surface, while the *Anopheles* lay eggs in "rafts"; eggs of both species hatch a few days after being laid.

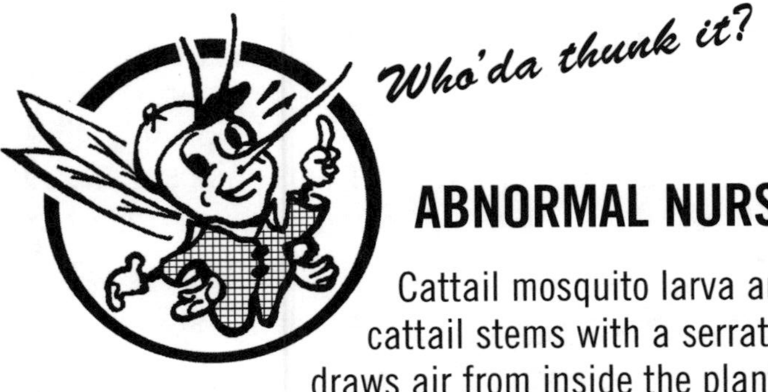

Who'da thunk it?

ABNORMAL NURSERIES

Cattail mosquito larva are attached to cattail stems with a serrated syphon, which draws air from inside the plant, reducing predation because they don't have to get air from the surface.

And if you think that's a strange place to raise a kid, the pitcher plant mosquito lays her eggs in the purple pitcher plant, a carnivorous plant that eats insects and spiders.*

***And sometimes even small frogs!**

DO MOSQUITOES HAVE A FAMILY LIFE?

Not really. As we said, mating is a one-night stand kind of affair. Although some tropical species defend their eggs to ward off egg-eating insects, they hardly get together for reunion picnics (unless, of course, it's to feed on human picnickers).

DO MOSQUITOES MIGRATE?

No—at least not intentionally. Some ride the wind until they are miles from where they were born.

Also, humans often unwittingly transfer species of mosquitoes across continents.

The Asian tiger mosquito was brought from a scrap tire pile in either Nagasaki or Kobe, Japan, to Houston, Texas.*

**Today, the Asian tiger mosquito is found in twenty-one states.*

Not so fast...

THEY GET AROUND...

Some species have been found 100 miles from where they were born, but most travel no farther than a mile or two from their birthplace.

Since they only fly about 2.5 m.p.h., who can blame them for not taking long trips?

IF THEY DON'T MIGRATE, WHERE DO MOSQUITOES GO IN THE WINTER?

Some adults mosquitoes in northern climates (if they don't first die of cold) spend the winters hiding out in barns, caves, tree holes, cellars, etc.—anywhere they can get out of the wind. To keep from freezing, they form glycerol, which acts as antifreeze.

Most species overwinter as eggs, lying dormant until hatching conditions are right.

A few experts hypothesize, however, that mosquitoes simply spend their winters at home watching TV and are only out in summer because of reruns.*

*Other researchers have deemed this theory "just plain silly."

WHAT DO MOSQUITOES DO ALL DAY?

Most spend their day hiding in dark, cool places such as in building cracks, in tree holes, or leafy bushes.

They only feed for an hour or so a day, but it may seem much longer because of the quantity of mosquitoes that live around you. Different species eat at different times, but most feed during the last two hours of sunlight.

HOW LONG DO MOSQUITOES LIVE?

Mosquitoes live an average of two weeks in the summer, but may live to be the ripe old age of about one month.*

Mosquito eggs can lie dormant for up to seven years waiting for ideal hatching conditions.

mosquito facts
Interesting info about skeeters

WHO CAME UP WITH THE NAME "MOSQUITO"?

In medieval England, a mosquito was called a "midge"— the word "mosquito" didn't appear until the sixteenth century. The etymology* of the word can be traced back to *musca*, the Latin word for "fly." This Latin word became *mosca* in Spanish and Portuguese. *Mosca* then became "little fly" or *mosquito* (also *mosquita*) to describe our favorite little pest. *Mosquito* was borrowed by English in about 1583.

Interestingly, the English word "musket" is also borrowed indirectly from Latin. The Latin *musca* and Italian *mosca* formed *moschetta*, meaning "bolt for a catapult" and "small artillery piece." From *moschetta* came *moschetto*, or "musket," the source of French *mousquet*.

And so "little fly" came to mean "bolt from a crossbow," appropriate enough since both crossbow projectiles and mosquitoes fly, buzz, and draw blood.

*"Etymology" (the study of words) and "entomology" (the study of bugs) are in no way derivatives of one another.

Wicked googily!

THE ENGLISH CALL MOSQUITOES "GNATS."

Of course, they also call trunks "boots," elevators "lifts," and trucks "lorries." Besides that, they drink their beer warm.

ONE FRENCH WORD FOR "MOSQUITO" CAN ALSO MEAN "COUSIN."

French: the perfect language for those who consider their relatives annoying little bloodsuckers.

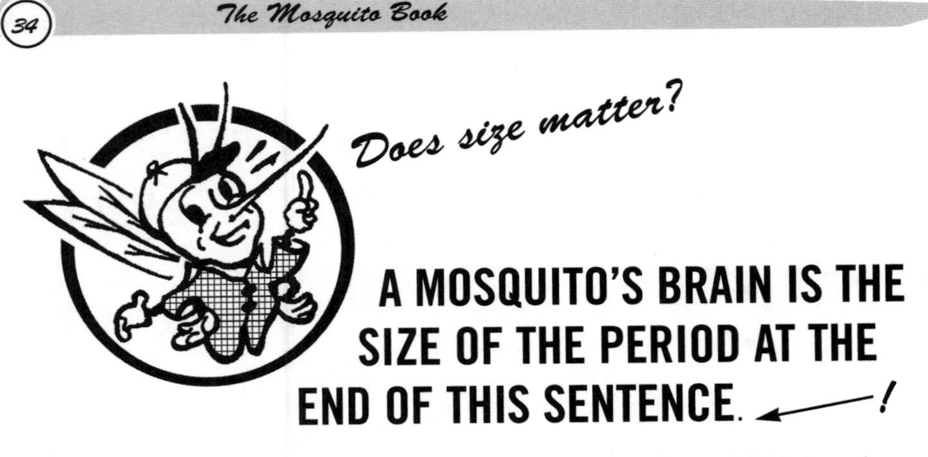

Does size matter?

A MOSQUITO'S BRAIN IS THE SIZE OF THE PERIOD AT THE END OF THIS SENTENCE. ←!

And yet they have outwitted mankind since the dawn of history.*

***Mosquitoes have been consuming blood for about 200 million years.**

WITH SUCH TINY BRAINS, HOW DO MOSQUITOES THINK?

Mosquitoes don't think, at least not in the conventional sense. Mosquitoes behave according to a set of fixed action patterns that have been ingrained into their nervous systems after millions of years of evolution.

HOW MUCH DOES A MOSQUITO WEIGH?

A typical mosquito weighs around 2.5 milligrams.

That's about twenty thousand mosquitoes to a pound.*

***Thus the reason recipes featuring mosquitoes have never become popular: too darn much prep work.**

MOSQUITOES PREY ON US, BUT WHAT PREYS ON MOSQUITOES?

Bats, birds, other insects, bacteria, fungus, lizards, spiders, fish, and (slap!) humans.

However, even though millions of mosquitoes become meals every day, their populations are not significantly reduced by predators.

HOW MANY MOSQUITOES ARE OUT THERE?

About 100 trillion mosquitoes from 3,450 different species* are ready to pester humans on any given day.

Assuming the bugs are 1/4" by 1/4" by 1/4", 100 trillion mosquitoes stacked together on a football field would create a pile over 3 miles high (assuming they didn't squash each other).

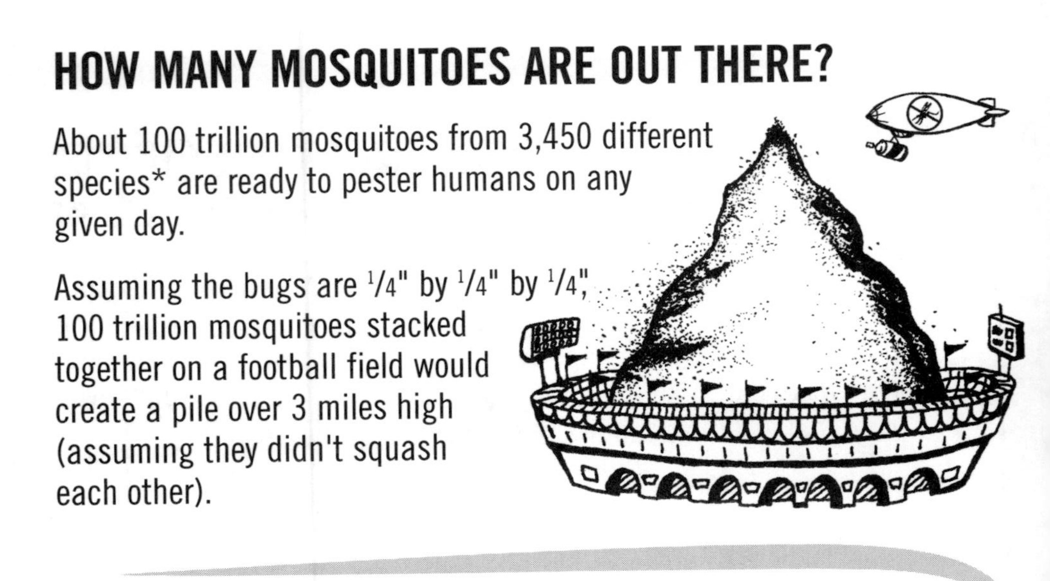

*Fortunately, the U.S. hosts only 170 species; Canada has roughly 70 more.

WHERE ARE MOST MOSQUITOES FOUND?

Wherever you happen to be wearing shorts in the summer.

And pretty much everywhere else.

Tropic regions have the most varieties of mosquito species, but the Arctic and Antarctic regions are home to larger populations of mosquitoes.

They have been found at elevations of 14,000 feet in the mountains of Kashmir and 3,800 feet below sea level in gold mines in Southern India.

Why would they do that?

9,000 BITES PER MINUTE!

In Canada, during an experiment when hordes of mosquitoes actually darkened the sky, researchers who allowed themselves to be exposed to mosquitoes were bitten about nine thousand times in one minute!*

At that rate, adult human victims would lose half of their blood in about two hours—enough to kill them!

**Some mosquito swarms in the Arctic are the size of small states!*

WHY DO SOME MOSQUITOES BUZZ LOUDER THAN OTHERS?

You may notice that mosquitoes seem to get louder when you are trying to sleep. They're not really louder: they're buzzing closer to your ears than they normally would because your head is the only thing not covered when sleeping.

This same theory can be applied to spousal snoring.

WHY IS IT SO HARD TO SWAT A MOSQUITO?

They may seem slow, but mosquitoes are tricky flyers: They can move up, down, sideways, or backwards.*

However, it is most likely that the wind created by your swatting hand is actually blowing the mosquitoes away and keeping you from swatting them successfully.

Some people claim mosquitoes can manuever around raindrops, but this has not been proven.

HOW DO MOSQUITOES FIND US?

Mosquitoes live in a chemical world—they change their flight pattern depending on what they smell.

Mosquitoes come a-callin' after bumping into chemical or physical clues coming from their potential victim: carbon dioxide, lactic acid, natural skin oils, and heat.

Basically, if you can prevent yourself from breathing and sweating, they'll leave you alone.

WHY DO MOSQUITOES GO AFTER SOME PEOPLE MORE THAN OTHERS?

Some people are more attractive to mosquitoes because they smell better (or worse, depending on your perspective) than others.*

*Unlike guard dogs, mosquitoes can't "smell your fear."

WHY DO MOSQUITOES SEEM TO LIKE HUMANS SO MUCH?

Actually, we are second choice for many species; some prefer birds.

We're also not the best meal for mosquitoes. Human blood is low in isoleucine, an amino acid mosquitoes need in order to build their egg proteins.

We are, however, easy prey—we're big and smelly, and there are a lot of us to bite. And the more we build, the more we force other animals to move, so we give the mosquitoes little choice but to chew on us.

WHY DO MOSQUITOES LIKE ANKLES SO MUCH?

We don't know, but our researcher, Laura Witrak, has a theory:

"In the summer you wear sandals; your feet stink a little more (which attracts mosquitoes), and the ankles are near your feet. Voila—that's my explanation."*

*Speak for your own feet, Laura.

WHY DO YOU GET "BUMPS" AFTER A MOSQUITO BITE?

Before sucking your blood, the female mosquito injects you with her saliva, which contains an anticoagulant that allows the blood to flow freely into her (this is also how she transmits disease).

The "bump" is your body's reaction to a protein contained in mosquito spit.

WHY DO PEOPLE REACT DIFFERENTLY TO BITES?

It all depends on your skin sensitivity and predisposition to allergic reactions.

It also depends on how many times you've been bitten in your lifetime.*

**Some researchers have been bitten so many times they no longer react to mosquito saliva.*

POISON!

You may have heard this before: "When you get bit, mosquitoes inject you with a poison—that's why you get an itchy bump. So wait until it's done feeding (and has sucked up the poison with your blood) before killing it."

Not true! There is no poison—just an anticoagulant (and, perhaps, a virus harbored in the saliva; see page 47).

ARE SOME PEOPLE NATURALLY IMMUNE TO MOSQUITO BITES?

Nope. People who have never been bitten by a mosquito will not react—except, perhaps, with a slap—to their first mosquito bite, but they are not immune.*

*It takes repeated bites before the skin reacts.

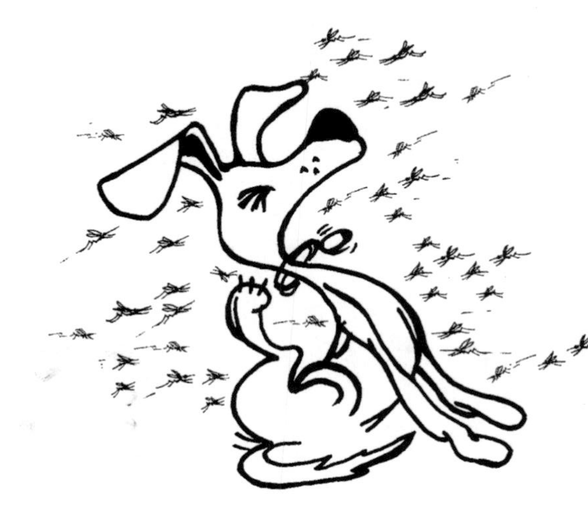

DO NONHUMAN MOSQUITO-BITE VICTIMS ITCH LIKE WE DO?

Yes! Horses, cattle, the family dog—mosquitoes make everything itch!

We humans, however, seem to be the only ones complaining about it.

ARE MOSQUITOES GOOD FOR ANYTHING?

Many species of mosquitoes help pollinate flowers.

They're also an important link in the food chain: they provide food for other insects and organisms that don't bother humans.*

They also provide work for entomologists, public health specialists, physicians, and insect-repellent manufacturers (not to mention mosquito-book authors, researchers, illustrators, designers, copy editors, proofreaders, and publishers).

*They also help diagnose disease—see page 68.

mosquitoes & disease
The truth and the myths

CAN MOSQUITOES TRANSMIT AIDS?

No!

While mosquitoes can spread more than one hundred viral diseases* and naturally ingest the HIV virus whenever they take blood from an infected person, the AIDS-causing virus dies quickly inside the mosquitoes' stomachs.

They can't carry the Ebola virus, either!

*Because of the diseases they spread, mosquitoes kill more humans than do any other animal.

A CAMEL-LOT OF SKEETERS

'Round about 597 B.C., Babylon was besieged by plagues.

Historians suspect that camel caravans—and the malaria-carrying mosquitoes that bred in polluted irrigation canals along the caravan route—were the cause.

MALARIA

Malaria affects 300 million people per year, and it is transferred to humans solely by mosquitoes. A disease of the tropical world, malaria is found in 102 countries. Only a few cases have been found in the United States since the 1950s.

Symptoms include high fever, chills, headache, fatigue, nausea, vomiting, diarrhea, anemia, spleen enlargement, liver and kidney failure, and finally brain damage, which can lead to death.

Malaria has no vaccine. Quinine (the same stuff that's in tonic water) was once used to cure malaria, but the disease has developed resistance to it.* In the 1960s, malaria was thought to have been wiped off the face of the earth. Insecticides were killing mosquitoes and drugs were curing the last cases. The mosquito has since developed resistance to insecticides and drugs, and malaria is back, baffling scientists as to how to stop it.

*"*Malaria prevention*" *is no longer a good excuse to drink gin & tonics.*

THE FALL OF ROME

In 395 A.D., 330,000 acres of farmland in Rome's Campania region were abandoned due in part to a malaria epidemic brought on by mosquitoes that bred in nearby swampy areas.

Rome fell just eighty-one years later. Coincidence? You be the judge.

FILARIASIS & DENGUE

Filariasis is caused by filarial nematodes transmitted by mosquitoes. This disease is found in sections of Asia and Africa where sanitation is poor. It affects 250 million people in fifty countries, but has not yet been found in the United States.

Symptoms include recurrent fevers, swollen lymph nodes, and swollen limbs. As the disease progresses, the limbs and genitalia become increasingly swollen.*

Dengue is a fever sickness caused by a mosquito-borne virus. This illness has been recorded in the Americas, Africa, and Asia. Worldwide, nearly one million people in fifty countries are afflicted with dengue. It has not been found in the United States since the 1920s, when 600,000 people in Texas were affected.

Symptoms include high fever, bone and joint pain, intense headache, skin rash, small hemorrhages, nausea, vomiting, swollen glands, fatigue, and depression.

***More easily amused readers can insert their own joke here.*

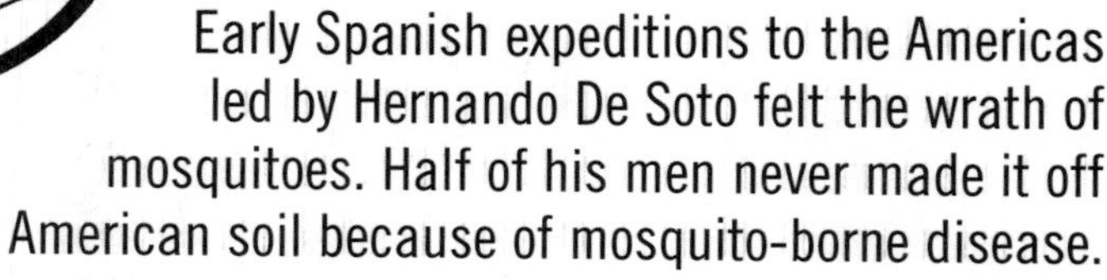

FOILING THE SPANISH

Early Spanish expeditions to the Americas led by Hernando De Soto felt the wrath of mosquitoes. Half of his men never made it off American soil because of mosquito-borne disease.

YELLOW FEVER

Yellow fever is another virus-driven disease transmitted by our friend the mosquito. It is rampant in monkey communities and has been transferred to humans as we continue to encroach on the wild.* This virus affects ten thousand people per year and is found in twenty countries. The virus moved from Africa to the Americas with the slave trade in the 1500s, but has not been found in the United States since 1905 (see page 65).

Symptoms include sudden-onset fever, headache, nausea, slowed pulse, reduced urine production, and low white-cell count. In advanced stages, victims bleed from the mouth and nose, vomit blood, become jaundiced, and form lesions in the liver, kidneys, and gastrointestinal tract.

Fortunately, there is an effective vaccine.

***Yellow fever has wiped out entire populations of monkeys in South and Central America.**

A MAN A PLAN A CANAL PANAMA

In 1902, U.S. Surgeon General W. C. Gorgas led a team of scientists (including Walter Reed) in the fight to eradicate mosquitoes in Panama to pave the way for the construction of the Panama Canal.

France abandoned an earlier attempt to build a canal through Panama after losing many workers to mosquito-borne yellow fever and malaria.

FILARIASIS, DENGUE, YELLOW FEVER: IF THEY'RE RARELY, IF EVER, FOUND IN THE U.S., SHOULD WE EVEN WORRY?

Just because they have not yet found their way to the United States doesn't mean we're safe from these diseases.

If global warming theories are correct and mean world temperatures rise just a few degrees, mosquitoes carrying these diseases could increase their range and spread into North America.*

Also, there is a low probability that an airline commuter could contract a mosquito-borne disease elsewhere, fly to the U.S., and infect others.

***And we were worried about communists and killer bees.**

PURCHASING LOUISIANA

In 1802, Napoleon sent thirty-three thousand men to Haiti to quell an uprising. Mosquito-borne yellow fever claimed twenty-nine thousand of them. As a result, the U.S. was able to land the Louisiana Purchase (Thomas Jefferson's vision of western expansion) because France had no holding power nearby.

Napoleon also needed the cash for his plans to invade England in flat-bottomed boats.

ENCEPHALITIS

The one mosquito-borne virus infectious to humans that has a stronghold in the United States is encephalitis.

This virus is passed back and forth between mosquitoes and either birds, chipmunks, or squirrels until an infected mosquito bites an amplifying host (an animal that does not get sick from the virus but rather helps the virus to mature and reproduce).

For example, a mosquito carrying immature encephalitis virus bites and infects a chipmunk. The virus matures in the chipmunk's blood. Later, another mosquito bites the same chipmunk and picks up the virus. The mosquito can now pass it on to humans, who experience horrible symptoms: sudden-onset fever, headache, stiff neck, brain inflammation, convulsions, and sometimes coma.*

**People who do survive these bouts become immune to the viruses.*

A HOT TIME IN NEW ORLEANS

In 1905 a U.S. Public Health Service anti-mosquito campaign put an end to an epidemic of yellow fever that killed at least one thousand people in New Orleans (it was the last epidemic of yellow fever in the United States).

HEARTWORM DISEASE

Most dog owners are well aware that mosquitoes transmit heartworm disease to dogs. This disease, a parasitic worm that can grow up to fourteen inches long and lives in a dog's pulmonary arteries, is found all over North America. There is no vaccine, but monthly doses of antiparasitics called milbemycin oxime and ivermectin prevent dogs from developing the disease by killing the worm while it's still in its larval stage.

If your dog is not taking a heartworm preventive medication, get that pup to the vet!*

**Cats can get it too, but it's not as common in felines as it is in dogs.*

HEROES OF THE MOSQUITO WARS

SIR PATRICK MANSON (1844–1922)

Englishman Manson first made the assertion that mosquitoes transmit malaria (1877).

SIR RONALD ROSS (1857–1932)

A British physician, Ross won a 1902 Nobel Prize for proving that mosquitoes transmit malaria.

CARLOS JUAN FINLAY (1833–1915)

This Cuban physician suggested in 1881 that the mosquito was the carrier of yellow fever and later specified the correct species, now known as *Aedes aegypti*.

WALTER REED (1851–1902)

An American army surgeon, Reed proved Finlay's theory of mosquitoes as the carrier of yellow fever. He also helped out with malaria problems in Panama and has a big ol' hospital named for him.

MOSQUITOES HELP DIAGNOSE DISEASE?

Through a procedure called xenodiagnosis, physicians can diagnose an infectious disease at an early stage by exposing a presumably infected individual or tissue to a clean, laboratory-bred mosquito and then examining the mosquito for the presence of the infective microorganism.*

***So they're good for something besides pollination, bolstering the food chain, and providing jobs.**

controlling mosquitoes
What works, what doesn't

THE DAWN OF MOSQUITO CONTROL

Humans have been trying to control mosquito populations for a long time and have dumped everything from oil of pennyroyal to crude oil in mosquito breeding grounds in an effort to get rid of them. Unfortunately, the crude oil also killed every other species in the water and ruined entire ecosystems.*

**Today, the U.S. and Canada spend about $150 million a year trying to control mosquito populations.*

DRAGONFLIES

Dragonflies, along with damselflies, eat lots of mosquito larvae and adults.

In fact, in the southern U.S., the dragonfly is also called the "mosquito hawk" or "skeeter hawk."

THE MOSQUITOFISH

Around the turn of the century, the mosquitofish (the most widespread freshwater fish in the world) was used to control mosquito populations. A little gray fish from the northeastern U.S., the mosquitofish is a rabid eater of mosquito larvae and can clear out an entire pond within days.

But they're not perfect. You see, in ponds with dense floating vegetation and organic debris, mosquitofish ignore the larvae. They also consume other beneficial insect larvae and larvae of fish that eat mosquito larvae. And they tend to take over new habitats and shove out other species.*

***For these and other reasons, this practice is not in widespread use today.**

PURPLE MARTINS

Although many people think that purple martins eat lots and lots of mosquitoes and that having them around will keep mosquito populations low, there is no biological evidence to support this claim.

Some experts believe this myth was spread by a certain "Dr. Wald" from Libertyville, Illinois, who claimed that a purple martin eats "6,600 mosquitoes a day."

No one knows where Dr. Wald got his information, but they do know his profession. Dr. Wald sold aluminum bird houses designed specifically for—you guessed it—purple martins!

DDT

The pesticide DDT (dichlordiphenylethylene) was once quite effective at controlling mosquitoes. It was also quite effective at destroying other animals, including many birds.

Fortunately, it is no longer used, in part because mosquitoes evolve at such a fast rate they have long since developed resistance to the chemical.*

DDT was also banned because it weakened the shells of bird eggs—especially those of eagles.

Unfortunately, humans evolve at such a slow rate we can't seem to find anything to effectively control mosquitoes.

BATS

Many people believe that because bats dine on insects, building a bat house on their property will keep mosquitoes away. They are wrong.

While bats eat lots of insects, mosquitoes make up just a fraction of their diet. Bats mostly enjoy beetles, moths, and leafhoppers. A recent study showed that mosquitoes made up only 0.7% of the stomach content in bats. Basically, bats eat whatever is handy, so you can't expect them to focus on just mosquitoes. And bats tend to migrate south in late summer, so there aren't many of them around when mosquito populations are at their peak.

MALATHION AND BTI: GMCA* WEAPONS OF CHOICE.

Today, many local governments control mosquitoes by spraying breeding areas with malathion, a short-lived chemical that is also, unfortunately, dangerous to beneficial insects.

Bti, the latest weapon, is a biopesticide designed to destroy mosquito larvae. Biodegradable within forty-eight hours, it contains bacterial spores whose toxic crystals destroy larval stomach linings, killing mosquitoes before they develop enough to get off the ground.

*Government Mosquito Control Agencies.

Smoke 'em if you got 'em!

THE BLACK CIGAR OF DEATH

In 1909, Brazilian rubber plantations stocked up on Il Negro Mortes, "the black cigars of death," in an effort to control mosquitoes. Workers earned extra cash smoking the cigars—which were made of tobacco, chemicals, and petroleum tar—on the verandas of plantation owners' mansions.

The fumes were effective, but deadly to all: it is estimated that one smoker died for every 5 billion mosquitoes that were killed.

TRAP 'EM!

Scientists on Florida's Key Island have experimented with mosquito traps that are essentially large barrels baited with carbon dioxide and a liquid octenol solution (which smells like cows' breath) and packed with a dose of a synthetic pyrethroid insecticide that destroys the mosquitoes on contact.

After a one-month test, the mosquito carnage from fifty-two traps filled three 30-quart coolers.*

***That's more than 2 billion mosquitoes!**

WHY TEST ON KEY ISLAND?

Because mosquitoes can be so thick there that if you pulled up your pant leg and waited fifteen to thirty seconds, your calf would turn black with five hundred feasting females.

In 1989, mosquitoes were so thick they were killing the island's cows: the bovines were inhaling so many mosquitoes they were choking and suffocating.

YARD AND HOME MOSQUITO CONTROL

Most mosquitoes breed in standing water, so do all you can do to reduce the standing water around your house:

- Keep grass cut short and shrubbery well trimmed around the house (mosquitoes love to hide in tall, cool grass and the shade of shrubs).
- Use sand to fill in any holes in trees or hollow stumps that hold water.
- Fill in holes and low spots in your yard (where water can build up).
- Empty your child's plastic wading pool and store it indoors when not in use.
- Change the water in bird baths and plant pots often.
- Keep drains and ditches free of debris (so water will drain).
- Cover trash containers.
- Remove any old tires, buckets, or empty planters.
- Repair leaky garden-hose faucets or any pipes that lead outdoors.*

**If all else fails, move to a drought-stricken area.*

repelling mosquitoes
What works, what doesn't

DEET: IT REALLY WORKS!

In the U.S. alone, 50 to 100 million cans of insect repellent are sold each year, so some of it must work, right?

Those containing the chemical DEET (N,N-diethyl-3-meta-tolumide) do, at least to some degree. DEET fools biting insects by masking our odors.

Repellents that contain DEET are by far the safest and most effective repellents currently on the market. However, too much DEET can cause eye and sinus irritability, headaches, insomnia, and confusion, so always remember to be careful of when and how you use it.*

***_And always read a repellent's label and follow its directions to the letter._**

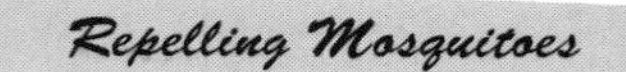

GOT DEET?

Always check containers of repellents that contain DEET to make sure you're using the safest amount for the conditions (see next page).

SOME POPULAR BRANDS OF REPELLENT AND THE PERCENTAGE OF DEET THEY CONTAIN:

Cutters™ Regular..........25.0%
Cutters™ Unscented..........21.85%
Cutters™ Pleasant Protection..........6.65%
Cutters™ Lotion (with sunblock)..........9.5%
OFF!™ Deep Woods Sportsman..........28.5%
OFF!™ DeepWoods..........25.0%
OFF!™ Skintastic..........7.5%
OFF!™ Skintastic with Aloe..........7.125%
OFF!™ Skintastic Kids..........4.75%

REPEL™ Sportsman (spray)..........16.62%
REPEL™ Sportsman (lotion)..........9.00%
REPEL™ Family (pump)..........16.62%
REPEL™ Family (aerosol)..........14.25%
REPEL™ Kids..........9.5%
Skeedaddle™..........6.5%
Nero™ (Available in Canada)..........75.0%
Natrapel™..........0%*

*Natrapel's active ingredient is 10% Citronella

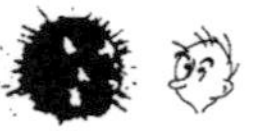

CAREFUL WITH THAT DEET!*

- Never use DEET on infants or on someone with open wounds.
- Children should use a repellent with NO MORE THAN 10% DEET.
- Adults need more than 10% only if they are in areas extremely thick with skeeters.
- Always wash DEET off after coming indoors.
- DEET can dissolve nylon and plastic.

*Even though it works, the less DEET you use, the better for your health.

AVON'S SKIN-SO-SOFT™

Does it work? Somewhat. It contains diisopropyl adapt benzophenome, which has a mosquito-repelling power, but is not as effective as DEET. Scientists consider it "unreliable."

Avon has never claimed that Skin-So-Soft™ repels mosquitoes, but they now market a repellent called Bug Guard.™

Now you big burly types have to come up with some other excuse to get baby-smooth skin.

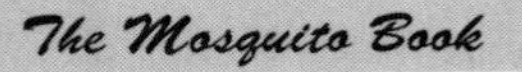

CITRONELLA CANDLES

While some studies claim people receive 42% fewer bites when burning a candle made with citronella oil (found in grass of China), these devices are most often ineffective. The candles give off chemicals that drive mosquitoes away, but only if there is no or little wind.*

***Never light one of these babies indoors—the chemicals that spook the skeets are also harmful to your health.**

THE CITROSA PLANT

The citrosa plant is a cross between the African geranium and grass of China (which contains citronella oil). Because the oils are emitted only when the plant is touched, the mosquito-repelling properties are rarely released.

Studies do not support the idea that the citrosa plant actually repels mosquitoes.

SPICY FOODS & VITAMIN B1

A popular myth is that eating lots of garlic and spicy foods will keep mosquitoes at bay.

Untrue!*

Some people also think taking vitamin B1 will make you naturally repellent to mosquitoes. The FDA considers this claim "unsupported."

***Garlic may ward off vampires, but tiny bloodsuckers like mosquitoes enjoy eating Italian.**

GERANIUMS

Venice, Italy, is infested with mosquitoes: all those canals mean lots of standing water.

Almost everywhere you look, window boxes are filled with geraniums.

Why? To ward off our buzzing buddies. Apparently geraniums harbor a citronella-like chemical. Research does not support geraniums as an effective repellent.

Venetians are apparently turning a blind eye to this fact.

CHRYSANTHEMUMS

Using chrysanthemums as a mosquito deterrent makes much more sense than using geraniums: pyrethrum, a derivative of which is used in propane foggers (see next page), is made from the extract of chrysanthemum.

Insect repellents such as Pet & Premises™ are designed for dogs and cats and include chrysanthemum extract because it is safe for the animals and indoor use. They generally cause insects to have spasms which knock them to the floor. Then you can just vacuum them up!*

**However, while pyrethrum has good "knockdown" capability, it only lasts for a few hours.*

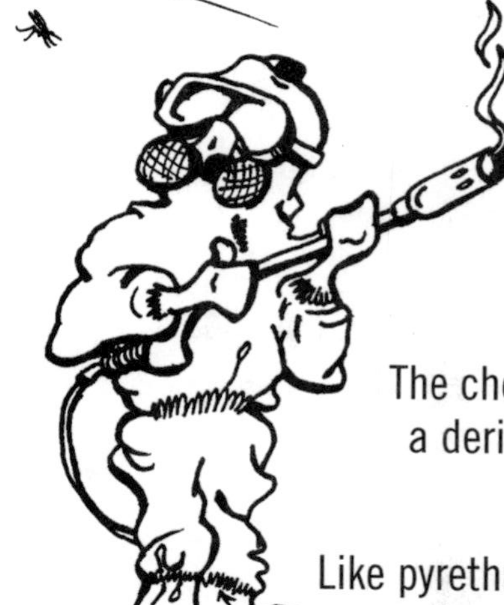

PROPANE FOGGERS

These work well in small areas, but they are not cost efficient: they are expensive, and their effects are only temporary.

The chemical used in these foggers is pyrethroid, a derivative of pyrethrum made from the extract of chrysanthemums.

Like pyrethrum, it causes mosquitoes to have spastic attacks. It lasts longest if sprayed on lawn shrubs.

MOSQUITO COILS

Mosquito coils are plant-derived devices that you set on fire. The smoke they give off contains mosquito-repelling chemicals.

What's the chemical?

Pyrethroid, the same chemical found in foggers.*

*As with citronella candles, never light a mosquito coil indoors!

SONIC DEVICES

You can now purchase electronic devices that emit a noise at a frequency designed to repel mosquitoes.

Careful with your cash—many scientists consider them a waste of money: at a cost of anywhere from five to forty dollars, sound devices are not effective enough to be considered sound investments.

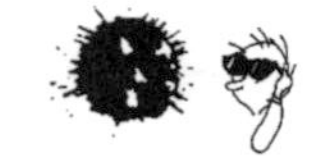

"BUG ZAPPERS"

Bug zappers (machines consisting of ultraviolet lights that attract insects and electric coils that fry them; they sell for about sixty dollars) do attract and kill insects. However, mosquitoes compose less than 7% of a catch, and only half of those are blood-feeding females. In fact, the zappers end up zapping a lot of insects that feed on more mosquitoes than the zapper can zap!*

Zappers can really backfire on you: Mosquitoes are attracted to ultraviolet light, but they are even more attracted to your smell. So when you set up a zapper in your yard, you end up attracting mosquitoes that, once they catch a whiff of human, turn from the zapper to you!

***Over 90% of all insects killed in zappers are beneficial insects.**

BOUNCE™ FABRIC SOFTENER?

Florida residents have been spotted wearing Bounce™ fabric softener sheets on their belts. Apparently they believe the scent keeps skeeters at bay.

A fashion faux pax? Not necessarily. Many residents of Florida are retired gentlemen over age 65: they have a tendency to wear white shoes and matching belts, so others rarely notice the fabric softener sheet.

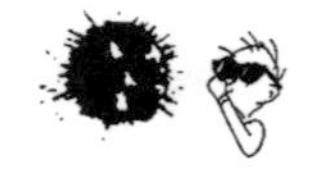

NATIVE AMERICAN REPELLENTS

According to outdoor writer Jeff Rennicke, Native Americans have traditionally used a variety of natural substances to deter mosquitoes, including deer fat, onion juice, cedar oil, and smoke smudge.*

The most popular Native American method for dealing with mosquitoes is also the simplest: ignore them.

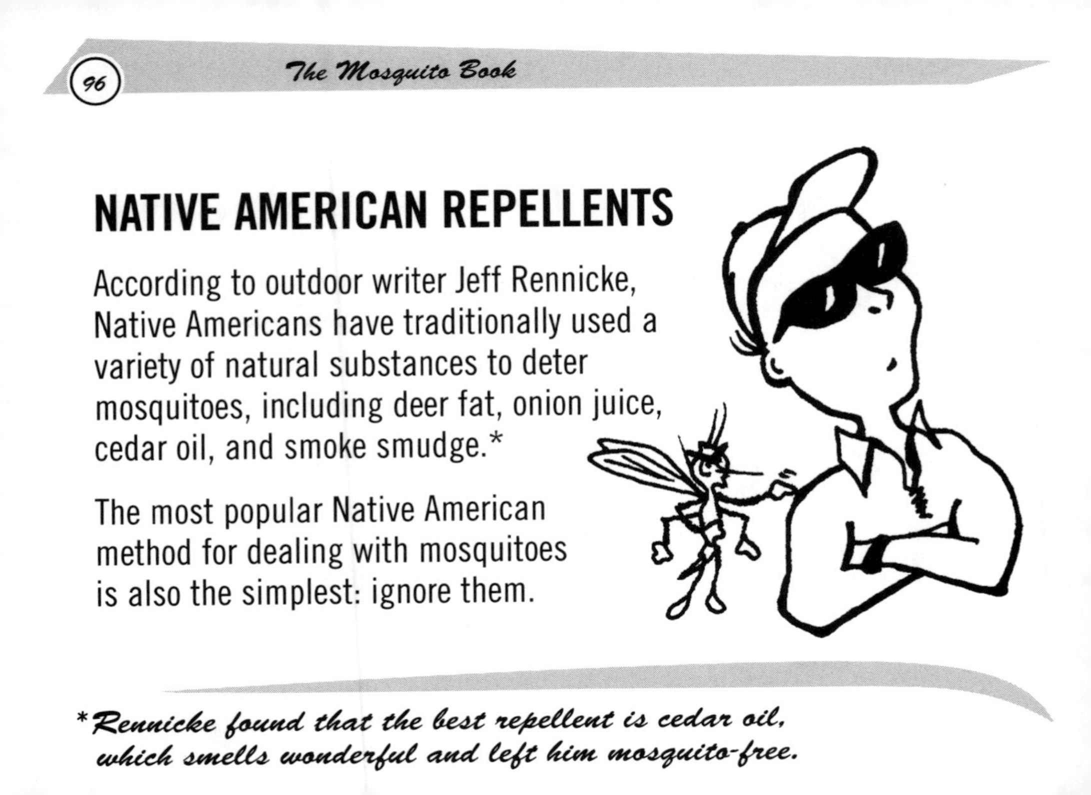

***Rennicke found that the best repellent is cedar oil, which smells wonderful and left him mosquito-free.**

STOP YOUR ITCHIN'!

Try these remedies for the allergic reaction that makes mosquito bites itch:

- In the early part of the century, lemon juice, vinegar, oil of peppermint, and oil of pennyroyal were used to help stop the itch.
- Rubbing alcohol effectively kills the sting but causes considerable pain. Washing the bite with soap and water can help.
- Icing the bump will reduce inflammation and swelling.
- Epsom salts and hot water work well together. One tablespoon of the salts dissolved in one quart of hot water, chilled, and applied to the bite, will help.
- Put the hottest water you can stand on a cloth and slap it on the bite. The itching should intensify briefly and then stop.
- Above all, don't scratch: it moves the saliva around and makes it itch worse.

PERSONAL MOSQUITO EVASION TACTICS

Want to do as much as you can to avoid mosquitoes?
Follow these common sense defense mechanisms:

- Wear loose clothing (the looseness provides a pocket of protection around you).
- Wear white or drab-colored clothing (mosquitoes are less attracted to dull colors). Never wear red—you might as well be ringing their dinner bell.
- Tuck pants into socks (so mosquitoes can't fly up your pants legs).
- Cover your head (wear a hat or buy a shirt with a hood).
- Stay inside during dusk and dawn (when mosquito feeding is at its peak).
- Do not wear a lot of smelly lotions, perfume, or scented hairsprays (remember, mosquitoes are attracted to smell).*
- Smoke from a campfire will keep mosquitoes away from the immediate area.

*If all else fails, wrap yourself in a huge plastic zipper bag.

mosquito miscellany
Tall tales and other tidbits

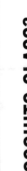

MOSQUITO CLASS?

According to *Backpacker* magazine, Teaching Drum Outdoor School in Wisconsin teaches a mosquito course where participants bare themselves "to Sister Mosquito in order to clearly hear her voice."

The course teaches students to revere the mosquito as a respected guardian of the North Woods. "In the Hoop of Life," says instructor Tamarack Song, "Mosquito is as vital and noble and beautiful as the Hawk or Grandfather Pine."

By the end of the class, campers conclude that the best ways to avoid Sister Mosquito is to "stay high and dry, avoid activity at dusk and dawn, stay out of the shadows and in the breeze, go slowly, wear green, and be first in line [when walking through the woods]."*

***Oh, and one more thing: "Don't breathe."**

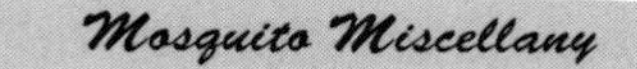

THE FOX AND THE MOSQUITOES
(ONE OF AESOP'S FABLES)

A fox, after crossing a river, got its tail entangled in a bush, and could not move. A number of mosquitoes, seeing its plight, settled upon it and enjoyed a good meal undisturbed by its tail. A hedgehog strolling by took pity upon the fox and went up to him:

"You are in a bad way, neighbor," said the hedgehog. "Shall I relieve you by driving off those mosquitoes who are sucking your blood?"

"Thank you, Master Hedgehog," said the fox. "But I would rather not."

"Why, how is that?" asked the hedgehog.

"Well, you see," was the answer, "these mosquitoes have had their fill; if you drive these away, others will come with fresh appetite and bleed me to death."

HOW MOSQUITOES CAME TO BE*

Long ago there was a giant who loved to kill humans, eat their flesh, and drink their blood. "Unless we get rid of this giant," people said, "none of us will be left." One man said, "I think I know how to kill the monster."

He went to the place where the giant had last been seen. There he lay down and pretended to be dead. Soon the giant came along. He touched the body and said, "Ah, good, this one is still warm and fresh."

The giant flung the man over his shoulder and carried him home. He dropped the man near the fireplace and went to get some firewood.

As soon as the giant had left, the man got up and grabbed the giant's knife. Just then the giant's son came in. He was still small as giants go, and the man held the big knife to his throat. "Where's your father's heart?" The giant's son was scared. He said, "My father's heart is in his left heel."

***Paraphrased from a Tlingit Indian legend.**

Just then the giant's left foot appeared in the entrance. The man plunged the knife into the heel. The giant screamed and fell down dead, yet he still spoke. "Though you killed me, I'm going to keep on eating you and all the other humans in the world forever!"

"That's what you think!" said the man. He cut the giant's body into pieces and burned each one in the fire. Then he took the ashes and threw them into the air for the winds to scatter.

Instantly each of the particles turned into a mosquito. The ashes became a cloud of mosquitoes, and from their midst the man heard the giant's voice laughing, saying, "Yes, I'll eat you people until the end of time." The man felt a sting, and a mosquito started sucking his blood, and then many mosquitoes stung him, and he began to scratch himself.

THE LION AND THE MOSQUITO*

Once upon a time a tiny mosquito started to buzz 'round a lion he met. "Go away!" grumbled the sleepy lion, smacking his own cheek in an attempt to drive the insect away. "Why should I?" demanded the mosquito. "You're king of the jungle, not of the air! I'll fly wherever I want and land wherever I please." And so saying, he tickled the lion's ear.

In the hope of crushing the insect, the lion boxed his own ears, but the mosquito slipped away from the now dazed lion. "I don't feel it any more. Either it's squashed or it's gone away."

But at that very moment, the irritating buzz began again, and the mosquito flew into the lion's nose. Wild with rage, the lion leapt to his hind legs and started to rain punches on his own nose. But the insect, safe inside, refused to budge. With a swollen nose and watery eyes, the lion gave a terrific sneeze, blasting the mosquito out.

**A very grim fairy tale from the Brothers Grimm.*

Angry at being dislodged so abruptly, the mosquito returned to the attack: BUZZ! BUZZ! It whizzed round the lion's head. Large and tough as the lion was, he could not rid himself of his tiny tormenter. This made him angrier still, and he roared fiercely. At the sound of his terrible voice, all the forest creatures fled in fear; but, paying no heed to the exhausted lion, the mosquito said triumphantly, "There you are, king of the jungle! Foiled by a tiny mosquito like me!"

And highly delighted with his victory, off he buzzed. But he did not notice a spider's web hanging close by, and soon he was turning and twisting, trying to escape from the trap set by a large spider. "Bah!" said the spider in disgust, as he ate it. "Another tiny mosquito. Not much to get excited about, but better than nothing. I was hoping for something more substantial...."

And that's what became of the mosquito that foiled the lion!

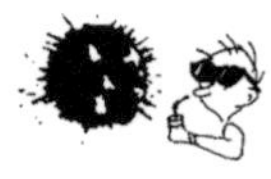

THE GREAT SHEEP & BUNNY WAR

In 1951, rabbits were overrunning sheep-herding land in Australia, eating enough grass to feed 40 million sheep! Sheepherders tried introducing hawks, weasels, snakes, and thousands of miles of anti-rabbit fencing, but that didn't stop these bunnies, who kept eating and eating and eating.

Finally, the government introduced the disease myxomatosis—a mosquito-borne virus—which brought the bunny population under control.*

In some areas 99% of the rabbits were eliminated.

MOSQUITOES AND GRUNGE ROCK?

Nirvana's breakthrough 1991 hit, "Smells Like Teen Spirit," opened the doors for feedback-happy northwestern bands who liked to play loud and wear plaid. That same song features the lyrics "I feel stupid and contagious, here we are now, entertain us, a mulatto, an albino, a mosquito, my libido, yay, yay, a denial...."

Any song with "smell" in the title and "mosquito" and "contagious" in the lyrics is thick with subtext: mosquitoes are most attracted to the way we smell and, as you know, they can carry encephalitis, yellow fever, and heartworm.

Also, the fuzz- and feedback-enhanced guitar work on many grunge songs has been described by some as an "annoying buzz."

Coincidence? Maybe. Or maybe we should just look to the last word of that song and the title of the album on which it appears: "Nevermind."

MOSQUITOES AND THE MOVIES #1

In the finest traditions of *Attack of the Killer Tomatoes*, the movie industry has blessed us with a film about really annoying mosquitoes.

Mosquito! was released in 1995 and received three and a half stars—out of a possible five—from the *1998 Blockbuster Entertainment Guide to Movies and Videos*, which used the following sentence to describe the movie:

"Above-average giant bugs-on-the loose flick pits campers and escaped convicts against human-sized skeeters that had mutated after drinking the blood of crash-landed aliens."*

***That's right, folks: three and a half stars.**

MOSQUITOES AND THE MOVIES #2

In the blockbuster *Jurassic Park*, scientists cloned dinosaurs using dino DNA found inside ancient mosquitoes trapped in amber. Could it work?

Odds are, not in our lifetime. The DNA will probably not be complete (as in the movie, where scientists supplemented the DNA with frog DNA), so it would be awfully tough to get it right.

Even if they could find complete DNA, the dinosaur egg would still need a place to develop. You see, even the sheep they're cloning in Scotland need a real live sheep to grow inside; the scientists would need a real live dinosaur in which the eggs could develop.

WHAT'S THE "MOSQUITO COAST"?

It's a region of eastern Nicaragua and northeast Honduras. It was a British colony from 1655 until 1860, when it became the Mosquito Kingdom. Nicaragua took over in 1894, and Honduras has controlled its share since 1960.*

**** *Mosquito Coast* is also the title of a movie starring Harrison Ford based on a novel of the same name by author Paul Theroux.**

SWAT TEAMS?

The Wonderlake Campground in Denali National Park, Alaska, holds an annual contest to see who can slap the most mosquitoes.

Not to be outdone, Pelkosenniemi, Finland, has played home to an annual "World Championship of Mosquito Killing" since 1992. Teams of anti-mosquito slappers swarm the mosquito-thick highlands surrounding the town while being cheered on by throngs of spectators. The record? A mere seven kills in five minutes. Organizers blame the small number of kills on the very crowd the event gathers: mosquitoes are drawn away from competitors by the smells of the fans.

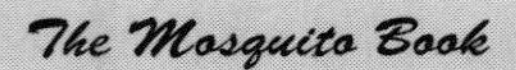

THE OFFICIAL MOSQUITO BOOK QUIZ

1. **Do male mosquitoes suck blood?** (Page 5)
   a. Yes.
   b. No.
   c. Only those who become personal injury attorneys.
2. **Which is required for a male mosquito to copulate?** (Page 19)
   a. His head must remain attached to his body at all times.
   b. His genitalia must first rotate 180 degrees.
   c. Candlelight and soft music.
3. **How fast can a mosquito fly?** (Page 27)
   a. 15 m.p.h.
   b. 2.5 m.p.h.
   c. Up to 75 m.p.h. in many states, but as fast as a prudent mosquito would fly in Montana.
4. **How many mosquitoes to the pound?** (Page 36)
   a. 50,000.
   b. 20,000.
   c. It depends; it takes twice as many if you use skinless, boneless mosquitoes.
5. **How did mosquitoes help Thomas Jefferson with the Louisiana Purchase?** (Page 63)
   a. France needed the money to pay for mosquito control in the sewers of Paris.
   b. They transmitted yellow fever to Napoleon's army in Haiti.
   c. They co-signed the loan to help Jefferson qualify for western-expansion financing.

6. **Which two scientists' work led to the discovery that mosquitoes transmit malaria?** (Page 67)
    a. Carlos Juan Finlay and Walter Reed.
    b. Sir Patrick Manson and Sir Ronald Ross.
    c. Juan Valdez and Walter Matthau.
7. **Which of the following is an effective, environmentally friendly form of mosquito control?** (Page 76)
    a. Saturate breeding grounds with DDT.
    b. Introduce the chemical Bti into mosquito-breeding sites.
    c. Carpet bomb breeding sights with napalm and thermonuclear ordnance.
8. **How can you best keep mosquitoes away from your house?** (Page 80)
    a. Build houses for bats and purple martins.
    b. Reduce the standing water around your property.
    c. Move without giving mosquitoes your new address.
9. **How does DEET work to repel mosquitoes?** (Page 82)
    a. It prevents mosquitoes from reproducing.
    b. It masks mosquito-attracting odors, confusing mosquitoes.
    c. It gives them a heightened sense of euphoria and the munchies, which repels them from you: they go to convenience stores for salty snack foods.
10. **How do mosquito-repelling sonic devices work?** (Page 93)
    a. Very well.
    b. They emit a noise at a frequency designed to repel mosquitoes.
    c. They play songs by Yanni, John Tesh, and the Spice Girls.

How do you think you did? Turn to the next page to check your score.

QUIZ ANSWER GUIDE

The correct answer to EVERY question is B. Give yourself one point for every correct answer and use the following guide to calculate your status:

0-3: Neophyte. You'll never survive in the North Woods. Reread this book and buy fifteen copies for friends.
4-6: Apprentice. You're in for a long summer. Reread book and buy ten copies for friends.
7-9: Entomologist. Use your score to bypass graduate school. Buy five copies for the university archives.
10: Honorary Minnesotan. Buy tons of copies and show your friends that you're better than they are.

EXTRA CREDIT!

Q: What 1960s sitcom featured an episode in which a Beatles-esque rock band called "The Mosquitoes" made an appearance? (bonus points if you know the names of all four Mosquitoes!)

A: *Gilligan's Island.* (John, Paul, George, and Ringo were parodied by actors playing Bingo, Bango, Bongo, and Irving. Incidentally, after Ginger, Mary Ann, and Mrs. Howell formed "The Honey Bees", the Mosquitoes became threatened and left the island—and our beloved castaways! Dohp! Gilligan!)*

***If you answered correctly, you obviously spent way too much time watching reruns when you should have been reading. Repeat grade 4.**

BONUS!

USE THIS BOOK TO REPEL AND EXTERMINATE MOSQUITOES!

When you've finished reading *The Mosquito Book*, put it to good use in your battle against mosquitoes!

The Mosquito Book Repellent:

By lighting this book on fire, you will produce smoke that will keep mosquitoes away from your immediate area as long as it is burning.

The Mosquito Book Extermination Device:

Take an old wooden spoon from the kitchen drawer, duct tape it to this book, and use your new tool to slap at skeeters while keeping your hands blood-free!

Bibliography

(Where we found our facts)

Adams, Sean. "A High-Tech Mosquito Barrier." *Agricultural Research.* (March 1996): 12-15.

Adler, Tina. "Mauling Mosquitoes Naturally." *Science News* 91 (1996): 270–272.

Aldhous, Peter. "Malaria: Focus on Mosquito Genes." *Science* 261 (1993): 546–548.

"All the Buzz." *Sports Illustrated*, 7/24/95, p. 14.

The American Heritage® Dictionary of the English Language, 3rd Edition. New York: Houghton Mifflin Company, 1992.

Barr, Ralph H. *The Mosquitoes of Minnesota*. Minneapolis: U of MN Experimental Station (Technical Bulletin 228), 1958.

Bates, Marston. *The Natural History of Mosquitoes.* New York: MacMillan, 1949.

Berenbaum, Mary. "The Natives Knew." *ChemTech* (May 1990): 275–279.

Bradshaw, William, and Christina Holzapfel. "Life in a Deathtrap." *Natural History* (July 1991): 35–36.

Carpenter, Stanley J., and Walter J. LaCrosse. *Mosquitoes of North America.* Berkeley, CA: University of California Press, 1955.

Child, Elizabeth. "Zen and the Art of Living with Mosquitoes." *Skyway News,* 5/15-21/97, p. 9.

Clancy, Frank. "The Mosquito Fighter's Survival Guide." *Health* 9, no. 4 (1995): 84.

The Columbia Dictionary of Quotations. New York: Columbia University Press, 1993.

The Concise Columbia Encyclopedia. New York: Columbia University Press, 1995.

Day, Jonathan F. "Epidemic Proportions." *Natural History* (July 1991): 50–53.

Edman, John D. "Biting the Hand That Feeds You." *Natural History* (July 1991): 8–10.

Erdoes, Richard, and Alfonso Ortiz. *American Indian Myths and Legends*. New York: Pantheon Books, 1984.

Gillette, Becky. "Controlling Mosquitoes Biologically." *BioScience* 38, no. 2 (1988): 80–81.

Gillette, J.D. *The Mosquito*. New York: Doubleday and Company, 1972.

Harwood, Robert F., and Maurice T. James. *Entomology in Human and Animal Health,* 7th Edition. New York: MacMillan, 1979.

Hawley, William A. "Adaptable Immigrant." *Natural History* (July 1991): 55–59.

Herms, W.B., and H.F. Gray. *Mosquito Control*. New York: The Commonwealth Fund, 1994.

Hertzberg, Hendrik. "Summer's Bloodsuckers." *Time,* 8/10/92, p. 46–48.

Klowden, Marc J. "Blood, Sex, and the Mosquito." *BioScience* 45, no. 5 (1995): 326–331.

——. "Tales of a Mosquito Psychologist." *Natural History* (July 1991): 48–50.

Lauerman, John F. "Mosquito with a Mission." *Health* (March 1991): 58–60.

Levin, Ted. "The Mosquito Coast." Sports Illustrated, 6/24/96, p. 5-8.

Mattingly, P. F. *The Biology of Mosquito-Borne Disease*. New York: American Elsener Publishing Co., 1969.

McCall, Bruce. "Five Historical Cigars." *Forbes*. 10/23/95, p. 182-183.

Nagel, Ronald L. "Malaria's Genetic Billiards Game." *Natural History* (July 1991): 59–61.

Nielson, Lewis T. "Mosquitoes Unlimited." *Natural History* (July 1991): 4–6.

Springer, Ilene. "Home Remedies that Really Work." *Ladies' Home Journal* (June 1993): 80.

Steel, Scott. "Summer's Sting." *MacLean's*, 6/26/95, p. 46–47.

Trager, James. *The People's Chronology*. New York: Henry Holt and Company, Inc., 1996.

Wilson, Samuel M. "Pandora's Bite." *Natural History* (July 1991): 26–29.

Index

(A fast way to find skeeter information)